Modeling Crop Responses to Irrigation
In Relation to Soils, Climate and Salinity

Modeling Crop Responses to Irrigation
In Relation to Soils, Climate and Salinity

R. J. Hanks and R. W. Hill

Department of Soil Science and Biometeorology, College of Agriculture, Utah State University, Logan, Utah, USA

International Irrigation Information Center

DISTRIBUTED BY PERGAMON PRESS

U.K.	Pergamon Press Ltd., Headington Hill Hall, Oxford OX3 0BW, England
U.S.A.	Pergamon Press Inc., Maxwell House, Fairview Park, Elmsford, New York 10523, U.S.A.
CANADA	Pergamon of Canada, Suite 104, 150 Consumers Road, Willowdale, Ontario M2J 1P9, Canada
AUSTRALIA	Pergamon Press (Aust.) Pty. Ltd., P.O. Box 544, Potts Points, N.S.W. 2011, Australia
FRANCE	Pergamon Press SARL, 24 rue des Ecoles, 75240 Paris, Cedex 05, France
FEDERAL REPUBLIC OF GERMANY	Pergamon Press GmbH, 6242 Kronberg-Taunus, Pferdstrasse 1, Federal Republic of Germany

Cataloguing in Publication Data
Hanks, R. J.
 Modeling crop responses to irrigation in relation
 to soils, climate and salinity.
 –(International Irrigation Information Center.
 IIIC publications; no. 6).
 1. Crop yields–Mathematical models
 2. Irrigation–Mathematical models
 3. Arid regions agriculture–Mathematical
 models
 I. Title   II. Hill, R. W.   III. Series
 631.5'58   SB51   80–40656

 ISBN   0–08–025513–2

Published by the International Irrigation Information Center, a non-profit corporation, whose activities are supported by the State of Israel and the International Development Research Centre, Canada
ISRAEL : P.O.B. 49, Bet Dagan
CANADA: P.O.B. 8500, Ottawa, K1G 3H9

Printed in Israel by Keterpress Enterprises, Jerusalem

Contents

Preface

The efficient management of irrigation is rapidly becoming a critical issue in agricultural development in arid and semiarid regions. The source of this urgency is the increasing cost of bringing water to the farmer's field and the dwindling of water resources in many parts of the world.

Sound irrigation management requires dependable information on water consumptive use and production functions, as well as on the dangers of increasing salinity with increasing efficiency. Such information is scarce and difficult to obtain experimentally at a specific location. A tool which will permit estimation of crop yield based on known climatic, soil and crop parameters is therefore highly desirable.

The purpose of the models discussed in this volume is the prediction of the crop yields under conditions of varying degrees of salinity and soil conditions, and under the various irrigation strategies available to the farmer.

The authors present a critical review of the evolution of yield models beginning with those based on evapotranspiration, leading to transpiration–available soil water models, and concluding with dynamic transpiration–soil water flow models. The latter include one-dimensional soil water movement, salt flow and salt accumulation, root distribution and extraction aspects, and rainfall/irrigation parameters.

Numerous examples of field measured data compared with calculations using the models are presented. They provide the reader with an understanding of the potentially practical applications of these models as well as the present limitations in their use.

This is an important book which can serve both as an introduction for scientists who are as yet unfamiliar with the subject and as a concise, comprehensive summary for those who have already been involved in related research. Instructors will find it useful as a basic text in university courses dealing with crop and irrigation management topics. In addition, irrigation advisors and extension service staff throughout the world will profit by this review of recent research in an area of utmost importance in irrigation planning.

Introduction

Models have been developed in recent years that solve complex problems by taking into account many factors and interactions. Several models that have been used to estimate crop yield responses to irrigation management are discussed. They range from simple models involving one or two equations to highly complex ones requiring large computers for solution.

The simplest model is the evapotranspiration (ET) model where crop yield is related to ET by a linear regression equation. Stewart's model is discussed as an example of this type. The model is somewhat site- and year-specific and requires measurements or estimates of ET and maximum yield. A major problem involves the amount of soil evaporation, E, which varies from year to year and from site to site. This approach has been found by many researchers to give good approximate predictions.

An extension of the ET model are the transpiration, T, models which are based on the strong linear relation with yield that has been reported by many observers. These models eliminate the problems mentioned above and are more readily transferable to different locations and different years because the E part of ET is corrected for. However, since it is almost impossible to differentiate E from T by field measurements, the transpiration models discussed require some method for this purpose.

A simple transpiration model devised by Hanks (1974) uses a field capacity type water balance method to predict T and E separately from inputs of potential transpiration (T_M), potential evaporation (E_M), and soil water storage properties. Relative yield is assumed to be equal to relative transpiration. A disadvantage is that estimates of potential yield are required from some other source. The input data of E_M and T_M reflect crop properties (crop cover). Thus, the model itself is not changed from one crop to the next. The water balance method is similar to others used successfully in many situations throughout the world, so it can be used as a basis for other yield predictions. Good estimates of dry matter yield as influenced by irrigation management have been demonstrated for many different situations. A copy of the computer program is included in the appendix. When the yield desired is not total dry matter, the approach used to consider relative transpiration during different growth stages has not always given good predictions without considerable "tuning" of the model.

A transpiration model developed by Kanemasu and associates uses climate-crop relation information to compute potential T and E. A water balance method similar to that of Hanks (1974) is used. However, rather complicated relations based on field measurements are used to predict absolute dry matter and grain yield of corn so that local measurements are not required. The yield predictions are limited to the crop tested and would need to be revised considerably for another crop. The limited tests made using this model indicate good predictions.

The water balance method used in the above models is limited in that upward flow and salinity effects are not accounted for. Childs and Hanks (1975) devised a model based on basic soil water flow equations which corrects the above limitations. Salinity, irrigation, and drainage (or upward flow) were simulated adequately with upward flow from a water table. However, this model requires much more computer time and input information than does the simple model of Hanks (1974). The same assumption of relative yield equal to relative T is made for both models.

Childs *et al.* (1977) have modified the transpiration model of Childs and Hanks (1975) to allow for computations of actual dry matter and grain yield at the expense, of course, of more computational time. This model was used to predict the influence of some limited irrigation practices on corn production, and gave good results. The yield prediction part of this model would need to be changed drastically for another crop.

List of Symbols

AC	Available carbohydrates in leaf
α	Crop and location coefficient
AW	Total available water of a given soil layer
A(z)	Root extraction term
b	Ratio of SWC/AW below which $T \neq T_M$
β_M	Coefficient relating yields to $ET_{D,V}$ in the maturity stage
β_P	Coefficient relating yield to $ET_{D,P}$ in the pollination growth stage
β_V	Coefficients relating yield to $ET_{D,M}$ in the vegetative growth stage
β_0	Slope of the relative yield *vs.* ET relation
c	Soil water transmitting coefficient for evaporation
C	Salt concentration
CU	Coefficient of uniformity
λ	Empirically determined growth stage constant
d	Growth stage weighting factors
D	Drainage
DM	Rooting depth
$D(\theta,q)$	Combined diffusion and dispersion coefficient — dependent on θ and q
DV	Average deviation from IA
ΔS	Change in soil water storage — the difference between the total amount of water stored in the soil at the beginning of a period and the end of the period
E	Evaporation from the soil
E_M	Evaporation from the soil (potential) where water is not limiting
E_O	Evaporation (potential) from a wet surface—equal to $T_M + E_M$ or equal to pan evaporation times an appropriate pan factor
ET	Evapotranspiration — sum of E from the soil and T from the crop
ET_D	Evapotranspiration deficit $= 1 - ET/ET_M$
ET_M	Evapotranspiration when water is not limiting crop growth ($T = T_M$). Note that E will normally be less than E_M
ET_{MA}	Potential evapotranspiration using alfalfa as a base
FC	Field capacity
$\Gamma(t)$	Relative growth in potential yield
γ	Psychrometer constant
h	Matric head
H	Hydraulic head
H_{Root}	Root water potential (head) at the soil

11

H_{Wilt} Limiting value of root water potential (head)

I   Irrigation

IA   Average irrigation rate

K_{CT} Crop coefficient related only to transpiration

$K(\theta)$ Hydraulic conductivity — dependent on θ

LAI   Leaf area index

LF   Lodging factor

m   Crop factor

P   Precipitation or rainfall

PH   Photosyntheses

ψ_l Leaf water potential

ψ_{1min} Minimum allowable leaf water potential

ψs Soil water potential

q   Water flux (volumetric)

R   Runoff

RDF   Root density function

R_n Net radiation

r_p Plant water flow resistance

r_s Soil water flow resistance $= 1/K(\theta)$

R_s Solar radiation

RS   Respiration

s   Osmotic head — related to soil solution concentration

S   Slope of saturation vapor pressure-temperature curve

σ_t Growth factor related to available soil water

SWC   Soil water content in a given soil layer

SYF   Seasonal yield factor

t   Time

T   Transpiration

τ Fraction of energy applied to soil surface

θ Soil water content (volume fraction)

t_M Maturity date

T_M Transpiration (potential) where water is not limited

t_R Time since beginning of reproductive phase

t_{RM} Total time in the reproductive phase

u   Soil water threshold coefficient for evaporation

Y   Yield of crop

Y_M Yield (potential) of crop where transpiration is never limited

Y_R Relative dry matter accumulation for the reproductive growth stage

Y_V Relative dry matter accumulation for the vegetative growth stage

z   Soil depth

Modeling Crop Responses
to Irrigation

Models of crop responses to irrigation have multiplied in the past few years. This has come about in part because of the need to look at a large number of factors that influence crop response to irrigation in a world of increasingly scarce and polluted water supplies. The response to this need has been facilitated by the widespread availability of computers and analysis techniques capable of handling complex systems.

A 'model' is another term for a set of equations that describe the physical system in question. These models can be very simple or very complicated and are, in reality, simply an organized expression of knowledge about the interacting factors in a given system. However, almost all systems are only partially understood, thus many simplifying assumptions are made to arrive at a practical model. A practical model is defined here as one that has been developed to the point that practical questions can be answered with some degree of reliability. We will concentrate on developed models which can predict crop yield response to soil water management (primarily related to irrigation). Many models have been and are being developed to predict plant physiological responses to various environmental factors (such as carbohydrate and nutrient balances, etc.). The only models considered here are those that allow prediction of ultimate yield as related to soil water conditions.

Present day models, of necessity, use simplifying assumptions to replace the details of plant response to the environment with less complex relationships. To include all the details of plant growth components would require larger computer programs and more understanding than we now have.

Evapotranspiration Models

The models which are of interest here are based on the assumption that crop production is directly related to evapotranspiration (ET) or evapotranspiration deficit (ET_D). Shalhevet et al. (1976) point out the strong relation between yield and ET for many crops. Stewart *et al.* (1977) have shown this for corn as have Hillel and Guron (1973). Stewart *et al.* (1975), Bielorai *et al.* (1964), and Hanks *et al.* (1969) have also shown a strong linear relation between grain sorghum yield and seasonal ET. Evapotranspiration is, of course, related to irrigation as well as other factors.

Stewart and co-workers (Stewart and Hagan, 1973; Stewart, Hagan, and Pruitt, 1974; Stewart, Misra, Pruitt, and Hagan, 1975) have used this approach. Their basic equation for dry matter production is

$$Y/Y_M = 1 - \beta_0\, ET_D = 1 - \beta_0 + \beta_0\, ET/ET_M \qquad [1]$$

where Y is actual dry matter yield, Y_M is maximum dry matter yield where $ET = ET_M$, ET is actual seasonal evapotranspiration, ET_M is maximum seasonal evapotranspiration, ET_D is evapotranspiration deficit and is equal to $1 - ET/ET_M$, and β_0 is the slope of the relative yield (Y/Y_M) *vs* the ET_D relation.

This equation requires that ET be measured or estimated. ET can be estimated by many procedures which will be discussed later. The water balance equation is often the basis for both estimates and measurements, and is given as:

$$ET = I + P + \Delta S - R - D \qquad [2]$$

where I is irrigation, P is precipitation, ΔS is the change in soil water storage, R is runoff, and D is drainage below the root zone.

ET_M will be determined by the above equation where growth is not limited by water (found where water content remains high during the season as ΔS is small). The problems associated with estimating or measuring ET using equation [2] arise in the measurement of R and D. To eliminate this difficulty, Stewart *et al.* (1974) used lysimeters which made R and D either zero or known. ET_M can also be estimated by procedures discussed later.

Note that the ratio ET/ET_M, where Y/Y_M is zero, can be shown to approximate the portion of ET_M that is due to evaporation (E) directly from the soil (Hanks, 1974). The portion of ET_M that is transpiration (T) is approximated by the fraction $(1 - 1/\beta_0)$. Thus β_0 should be 1 or greater. A value for β_0 of 1.0 would mean no evaporation from the soil and a β_0 of 1.5 would mean that one-third of E_M was E and two-thirds T. Measured values of β_0 less than 1 probably result from poorly defined values of ET_M or field variation.

Equation [1] together with [2] is quite simple, so unsophisticated calculation facilities can be used. However, for prediction purposes, a model using this approach would be more complicated because some scheme to estimate ΔS, R, D would have to be devised in response to a particular climate (precipitation) and irrigation regime. Such a scheme will be discussed later.

To illustrate the usefulness of this model, data from Stewart *et al.* (1977) are shown in Table 1 giving the results of a four-state study conducted over a two-year period for corn. Note that the value of β_0 varied from only about 1 to 1.3 for all locations, even though Y_M varied widely as did ET_M due to different local conditions. This illustrates that the coefficient β_0 is not strongly site- or year-related, which would imply that this approach is transferable to another location. It also indicates the usefulness of using relative yields and relative evapotranspiration data because the site-related variables are thus minimized. However, site specific measurements of Y_M and ET_M are required.

Stewart *et al.* (1977) have outlined a procedure whereby ET can be estimated in response to irrigation (or rain) using equation [2] but breaking it up into periods when water is applied. Measurements of soil water depletion, irrigation, and ET_M are needed for each period. Drainage is estimated assuming that a value of "available water holding capacity" of the soil is known. This is estimated from soil water measurements after thorough wetting and after maximum root extraction (or it may be estimated as the difference between field capacity and wilting water contents). Drainage is assumed to occur if the soil water storage capacity at the beginning of the period is not sufficient to contain the water added.

To illustrate the computation of yield using the Stewart model, the data from Stewart *et al.* (1977) can be used. For Logan, Utah in 1975, the measured ET of 39.8 and 47.0 cm, for two plots, gave predicted yields of 9.7 and 12.1 kg/ha respectively, using the data shown in Table 1. Measured yields for the same two plots were 10.3 and 13.8, respectively.

Table 1. Values of β_0, Y_M (dry matter), correlation coefficient, R^2, ET_M for different locations, years, and corn varieties as reported by Stewart *et al.* (1977)

Location	Year	R^2	β_0	Y_M, t/ha	ET_M, cm	Variety	Number of data
Davis,	1974	0.78	1.00	21.8	67.4	F4444	52
California	1975	0.56	1.00	22.8	61.6	F4444	37
Ft. Collins,	1974	0.64	1.09	18.0	52.9	NKPX20	23
Colorado	1975	0.72	0.92	16.2	54.5	P3955	25
Logan,	1974	0.72	1.16	14.4	64.4	UH544A	25
Utah	1975	0.71	1.23	15.2	56.2	NKPX20	27
Yuma,	1974	0.90	1.15	11.0	85.7	ASX504	40
Arizona	1975	0.66	1.29	10.4	67.4	RX450A	40

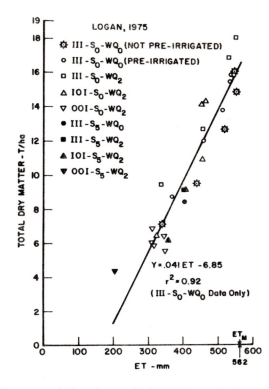

Figure 1.   Corn dry matter yields at Logan, Utah in 1975 as related to evapotranspiration, ET, irrigation timing and salinity. Treatment III indicates a series of treatments where irrigation was applied at all growth stages in varying amounts. Treatment IOI indicates that variable irrigation was applied at growth stage 1 (vegetative) and 3 (maturation), but not 2 (pollination), and OOI indicates that irrigation was applied only at growth stage 3. Salinity variables were S_0 (check) and S_5 (preplanting salinization of soil), and WQ_0 (check irrigation) and WQ_2 (saline irrigation).

The basic assumptions are also illustrated for the Stewart model by Figure 1, which shows dry matter yield as related to ET for corn for Logan in 1975 (Stewart *et al.*, 1977). Note that the data generally fit the model well even though there were growth stages where irrigation was applied and salinity variables were imposed which influenced yield as well as ET. Modification of the model to include irrigation–growth stage effects will be discussed later.

The question might be asked — "Why not develop a model where yield is related directly to irrigation and rain?" Many people have used such a simplification, but the end result is not always satisfactory, particularly when transfer of the results to other sites, other years or even different planting dates is desirable. This equation can be answered from an analysis of equations [1] and [2]. The water used by the crop comes not only from irrigation or rain, but also from soil water storage. Data from Davis, California (Stewart *et al.*,

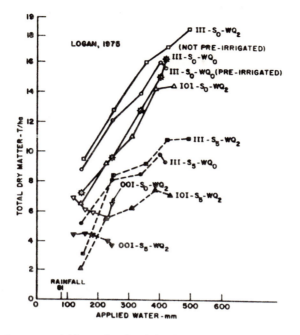

Figure 2.   Corn dry matter yields as related to irrigation applied at Logan, Utah in 1975. Treatments are as described in Figure 1.

1977) indicate that soil water depletion may be as high as 40 cm which was 59% of ET_M. Another problem in relating yields directly to irrigation is that part of the irrigation water may not be used by the crop but may be lost as drainage or runoff or be stored in the soil. Figure 2 shows a plot of dry matter yield *vs* applied water (irrigation and rain) for the same data for Logan in 1975 as shown in Figure 1. The data indicate large influences of the treatment on the results. For the OOI treatment (not irrigated until the maturity stage) irrigation had no effect on yields because the water was stored in the soil and not used. The salinity treatments also had much lower yields for the same irrigation. For these treatments soil water depletion was less, so that ET was less. Figure 1 shows that the yield *vs* ET data for the salinity treatments fit the same relation as the check treatment.

The model of Stewart (equation [1]) is probably the simplest that is practical and would seem to be useful for many situations where timing of irrigation is not of practical importance. Where total dry matter is the desired product, like alfalfa and corn silage, this model can be applied directly. However, for grain production, irrigation timing (related to growth stages) may be very important and the simple relation accounting only for seasonal ET may not predict well. Therefore, more complex models have been developed to account for different growth stage effects. However, even for grain the simple model may predict fairly well.

Stewart (as reported in Stewart et al., 1977) has also developed a model to predict the influence of the relative evapotranspiration in different growth stages on yield. The basic equation for grain production of corn is

$$Y/Y_M = 1 - \frac{(\beta_V \, ET_{D,V} + \beta_P \, ET_{D,P} + \beta_M \, ET_{D,M})}{ET_M} \qquad [3]$$

where β_V, β_P and β_M are coefficients for the vegetative, pollination and maturity stages, respectively. $ET_{D,V}$, $ET_{D,P}$, $ET_{D,M}$ are evapotranspiration deficits for the vegetative, pollination and maturity stages, respectively.

The values of $ET_{D,V}$ etc. were measured from data collected during the season. The β values were determined from a regression analysis of the data (Table 2) collected at various locations (Stewart et al., 1977). The data show a more complex situation because of "conditioning"—the effect of water deficits at a growth stage is conditioned by what happened previously. Thus, for a deficit occurring in the pollination stage, yield would be influenced to a different extent if a deficit had occurred in the vegetative stage than if there had been no deficit in the vegetative stage. The data of Table 2 indicate that the coefficient for corn in the pollination stage of growth is more sensitive than for the other stages of growth (a higher value) and is influenced by the "conditioning" phenomena. A major part of the treatments reported by Stewart et al. (1977) had water limitations throughout the season, so the growth stage effect was not too important. Thus a simple model like equation [1], with the β_0 coefficients for grain shown in Table 2, fits the measured grain yield data almost as well as does the more complicated equation [3]. However, many other

Table 2. Growth stage coefficients for the vegetative (β_V), pollination (β_P), and maturity (β_M) growth stages found for grain corn at different locations as reported by Stewart et al. (1977)

Location	β_V	β_P	β_M	O
Davis, California 1974 & 1975	1.3	2.8	1.1	1.1
Variety F4444		0.7*	1.1*	
Ft. Collins, Colorado 1975	0.8	1.1*	1.5*	1.0
Variety P3955				
Logan, Utah 1975	1.0	2.3*	1.2	1.4
Variety PX20			1.2*	

* Indicates conditioning of plant at earlier stage.

experiments, such as those reported by Stewart *et al.* (1975), have shown the importance of timing of water deficits. A general need exists for more complete models to account for these effects. This more complicated model is highly dependent on the crop and probably, to a lesser extent, on the variety of the crop.

Transpiration—Available Water Models

Other models have been developed that are more complicated than those of Stewart discussed above but are more general and sometimes allow for a more flexible predictive capability. Models relating yield to transpiration are more sound than those relating yield to evapotranspiration because they account for the water that goes through the plant. However, it is difficult to separate the two processes.

These models also predict evaporation and transpiration using a simple "field capacity — available water" treatment of soil water flow. Soil water flow is assumed to occur downward only and is also assumed to reach field capacity within one day. The influence of salinity on water availability is not considered.

Hanks Model

The model of Hanks (1974) is based on the evaluation of de Wit (1958) which in turn is based on the work of Briggs and Shantz (1913) and others. De Wit (1958) concluded that in the semi-arid parts of the world where irrigation was practised, there was a relationship between seasonal transpiration, T, and yield, Y, as follows:

$$Y = \frac{mT}{E_0} \qquad\qquad [4]$$

where E_0 is average seasonal free water evaporation and m is a crop factor. Figure 3 shows a plot of Y *vs* T/E_0 indicating a very good fit for several crops grown in different years and in different locations. These data, however, were all for an artificial situation because soil water was maintained at or near field capacity and evaporation from the soil was minimized because the plants were grown in small containers. Nevertheless, when the data were collected in the field and where soil water limited transpiration the results were similar. De Wit (1958) and Hanks (1974) also examined data where soil water was limiting and found equation [4] to represent the results well. Childs and Hanks

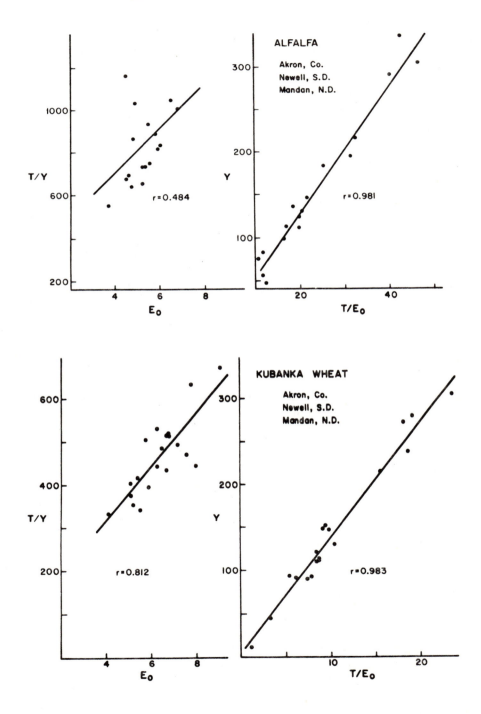

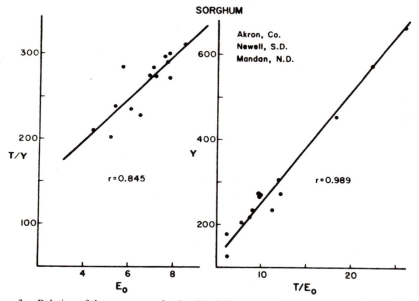

Figure 3.   Relation of dry matter production Y (yield) of alfalfa, wheat, and sorghum as related to transpiration, T, and free water evaporation, E_0, as discussed by de Wit (1958). Note the higher correlation if production is plotted as a function of T/E_0 rather than plotting T/Y vs E_0.

(1975) investigated the situation where transpiration was limited because of salinity and found the influence of yield to transpiration to be linear also.

Equation [4] has a disadvantage when used in practical situations in that a measure, or an estimate, of transpiration is needed, whereas evaporation plus transpiration are measured together in the field. This is why Stewart chose to use evapotranspiration as his main variable in spite of the problem of relating yield to evaporation as well as to transpiration. Hanks *et al.* (1969) have shown results where ET has been divided up into E and T under field conditions. The model of Hanks (1974) uses the approach that production should be based on estimates of T even though methods of estimating the separation of E and T are difficult. This is also the approach followed by Kanemasu *et al.* (1976). Van Keulen (1975) has used equation [4] as a basis for a crop growth model but has not applied it to irrigation situations.

In equation [4] the value of E_0 is a climatic variable for a season. If the value of Y is to be predicted as irrigation is varied (which may cause T to vary) at the same location for the same crop and year, another equation can be derived as follows:

$$Y/Y_M = T/T_M \qquad\qquad [5]$$

where Y_M is the maximum yield attained when soil water does not limit transpiration ($T = T_M$). This equation is analogous to equation [1] and is simpler because there is no intercept (where yield is zero and ET is not equal to zero). For the Hanks (1974) model where irrigation frequency is the same for all treatments, transpiration starts where the plot of ET/ET_M *vs* Y/Y_M intersects the ET/ET_M axis. This gives a method for estimating E and T from measured ET data. This approach only works, however, if all the treatments have been irrigated at the same time with enough water so that E is the same for all treatments.

Jensen (1968) has proposed a corn grain yield equation, which accounts for the effect of growth stages, similar to the following:

$$Y/Y_M(\text{grain}) = (T/T_M)_1^{\lambda_1} \cdot (T/T_M)_2^{\lambda_2} \cdot (T/T_M)_3^{\lambda_3} \cdot$$

$$\cdot (T/T_M)_4^{\lambda_4} \cdot (T/T_M)_5^{\lambda_5} \qquad\qquad [6]$$

where the "λ" values are weighting factors for the various growth stages and 1 through 5 are growth stage indices. Hanks (1974) has adopted this approach. The values of λ are empirically determined. Equation [6] does not generally give as good results for grain as equation [5] does for dry matter. Thus, a better equation for grain that has a sounder physical basis than equation [6] is being sought.

Hanks (1974) also devised a method for predicting T and E from soil plant climatic and irrigation inputs. It is a water balance model based on equation [2] with some refinements to partition ET into E and T. Generally the model requires input of soils, plant, climatic and irrigation data throughout the season and predicts T, T_M, E and E_M on a day-to-day basis. Thus, T/T_M can be calculated by growth stages at the end of the season as well as for the entire season, thus allowing prediction of yields. Table 3 illustrates the input data needed for this model as well as the day-to-day calculations.

The climatic data required are the values of E_0 by day or by periods and the amount and the time of rain. Other climatic data that have been used to

Table 3. Sample input data needed to run the model of Hanks (1974) for corn at Logan, Utah 1975. Day "0" is day of planting (May 28)

Climatic Data-Rainfall

Day	Amount, cm	Day	Amount, cm	Day	Amount, cm	Day	Amount, cm
6	5.80	10	0.20	11	0.79	21	1.63
22	0.23	24	1.04	28	0.33	62	1.29
63	0.13	64	0.84	79	0.56	105	0.30
111	0.30						

Period, days	E_0, cm/day	T_M, cm/day	Period, days	E_0, cm/day	T_M, cm/day	Period, days	E_0, cm/day	T_M, cm/day
0–8	0.24	0.0	9–16	0.62	0.0	17–23	0.44	0.01
24–30	0.54	0.08	31–37	0.69	0.23	38–44	0.73	0.35
45–51	0.67	0.43	52–58	0.63	0.48	59–65	0.66	0.55
66–72	0.72	0.60	73–78	0.54	0.45	79–85	0.58	0.48
86–92	0.60	0.50	93–99	0.58	0.48	100–106	0.41	0.34
107–117	0.41	0.33	118–128	0.41	0.32			

Irrigation Data

Day	Amount, cm	Day	Amount, cm	Day	Amount, cm	Day	Amount, cm
35	1.9	42	2.1	49	1.8	55	1.4
62	1.4	70	1.4	76	2.0	82	1.5
89	1.6	98	1.3				

Crop Data-Phenology (growth stage)

	Stage 1	Stage 2	Stage 3	Stage 4	Stage 5
Time (days)	0–20	21–63	64–75	76–87	88–128
λ	0.0	0.4	0.4	0.4	0.0

Root Growth

Date of maximum root growth, $RD = 90$ days

Maximum depth of root zone, $DM = 180$ cm

Depth roots $= DM/(1 + \exp(6-12 \cdot t/RD))$ where time, t, is in days

Soils-Crop-Available Water (AW)

Depth, cm	AW, cm	Depth, cm	AW, cm	Depth, cm	AW, cm	Depth, cm	AW, cm
0–45	4.5	46–90	5.8	91–120	4.2	121–150	3.9
151–180	2.1						

Soils Data-Beginning Soil Water Content (SWC)

Depth, cm	SWC, cm	Depth, cm	SWC, cm	Depth, cm	SWC, cm	Depth, cm	SWC, cm
0–45	4.5	46–90	5.8	91–120	4.2	121–150	3.9
151–180	2.1						

Air dry water content in top layer $= 2.0$ cm

define growth stages (using the growing degree day approach) are daily maximum and minimum temperatures (Rasmussen and Hanks, 1978, and Hill et al., 1974). The value of E_0 has generally been taken to equal Class A pan evaporation modified by an appropriate pan factor, although other climate methods could be used for this estimation.

The irrigation data required are the date and amount of irrigation that entered the soil.

The crop factors required are the rooting depth (DM) and potential transpiration (T_M) as a function of time as well as of phenology (growth stage timing), and the value of the λs. For some crops a static root zone is assumed (alfalfa) and for others a growing root zone is assumed (corn). Table 3 shows the sigmoid equation used to estimate rooting depth. Also needed is some information to use as a basis for estimating daily T_M separately from E_0. This is done, for a crop like corn, by starting with the crop coefficient data (ET_M/E_0) which are available from many sources (Doorenbos and Pruitt, 1977, or Jensen, 1973). The value of T_M is assumed to be equal to zero from planting until the start of growth, and is taken to be slightly less than ET_M (by about 10%) for the rest of the season. Figure 4 shows this estimate for corn as reported by Stewart et al. (1977) (see also Childs and Hanks, 1975). Also shown is the value of E_M which is computed from the following equation:

$$E_M = E_0 - T_M \qquad [7]$$

The soil factors needed as input data are the water holding capacity of each soil layer, the initial soil water content of each layer, and the air dry water content of the top layer. As reported by Stewart et al. (1977) these are best determined from field measurements of water depletion from a wet initial condition (field capacity) to a dry (wilting) condition. This can of course also be estimated by other means, such as from soil sample estimates of field capacity and wilting. However, for the same soil these values (field capacity minus wilting) do not appear to hold for all depths, especially near the bottom of the root zone. Thus, this input information is dependent on the crop as well as on the soil.

The relation needed to estimate T as influenced by soil water content, SWC, and potential transpiration, T_M (Hanks, 1974) is:

$$T = \frac{T_M}{b} \cdot \frac{SWC}{AW} \quad \text{if} \quad \frac{SWC}{AW} < b \qquad [8]$$

or

$$T = T_M \quad \text{if} \quad \frac{SWC}{AW} > b$$

where AW is the total available soil water.

This relation is similar to the estimates of many investigators as summarized by Tanner (1967). Hanks (1974) has used a value of $b = 0.5$. He divided the root zone into several layers and applied equation [8] to each layer. This allows for $T = T_M$, if one layer is wet and the others dry as is often the case following small rains or irrigations. Details of the computation are shown in Table 4.

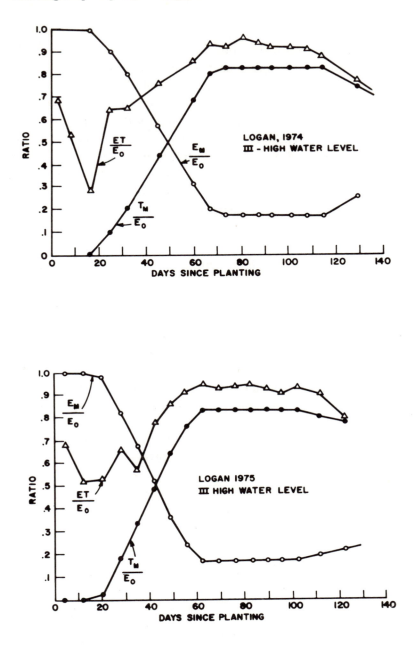

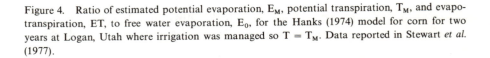

Figure 4.   Ratio of estimated potential evaporation, E_M, potential transpiration, T_M, and evapo-transpiration, ET, to free water evaporation, E_0, for the Hanks (1974) model for corn for two years at Logan, Utah where irrigation was managed so $T = T_M$. Data reported in Stewart *et al.* (1977).

Table 4. Example of daily computations of evaporation (E), and transpiration (T), for corn as related to irrigation, (I), climate (evaporation, E_0, from free water $\simeq T_M + E_M$, and rainfall, P) and soil (soil water content at different depths, $SWC_1 - SWC_5$) variables. Crop variables are also involved through the division of E_0 into T_M and E_M as well as rooting depth

Day	ET_M, cm	T_M, cm	E_M, cm	T, cm	E, cm	I, cm	P, cm	D, cm	SWC_1, cm	SWC_2, cm	SWC_3, cm	SWC_4, cm	SWC_5, cm
0									4.50	5.85	4.20	3.90	2.10
1	0.24	0.00	0.24	0	0.24	0	0	0	4.26	5.85	4.20	3.90	2.10
2	0.17	0.00	0.24	0	0.17	0	0	0	4.09	5.85	4.20	3.90	2.10
3	0.14	0.00	0.24	0	0.14	0	0	0	3.95	5.85	4.20	3.90	2.10
4	0.12	0.00	0.24	0	0.12	0	0	0	3.83	5.85	4.20	3.90	2.10
5	0.11	0.00	0.24	0	0.11	0	0	0	3.72	5.85	4.20	3.90	2.10
6	0.24	0.00	0.24	0	0.24	0	5.80	5.0	4.26	5.85	4.20	3.90	2.10
										a			
40	0.40	0.35	0.38	0.35	0.15	0	0	0	0.12	4.80	4.20	3.90	2.10
41	0.40	0.35	0.38	0.35	0.15	0	0	0	−0.02	4.45	4.20	3.90	2.10
42	0.73	0.35	0.38	0.35	0.38	2.10	0	0	1.70	4.10	4.20	3.90	2.10
43	0.62	0.35	0.38	0.35	0.27	0	0	0	1.43	3.75	4.20	3.90	2.10
44	0.56	0.35	0.38	0.35	0.21	0	0	0	1.21	3.40	4.20	3.90	2.10
45	0.55	0.43	0.24	0.43	0.12	0	0	0	1.09	2.97	4.20	3.90	2.10
46	0.54	0.43	0.24	0.43	0.11	0	0	0	0.98	2.97	3.77	3.90	2.10
											a		
77	0.52	0.45	0.09	0.45	0.07	0	0	0	2.16	1.78	1.02	0.69	0.16
78	0.50	0.45	0.09	0.44	0.05	0	0	0	1.66	1.78	1.02	0.69	0.16
79	0.58	0.48	0.10	0.47	0.10	0	0.56	0	1.65	1.78	1.02	0.69	0.16
80	0.55	0.48	0.10	0.43	0.07	0	0	0	1.24	1.69	1.02	0.69	0.16
81	0.54	0.48	0.10	0.37	0.06	0	0	0	1.18	1.41	0.92	0.69	0.16
82	0.58	0.48	0.10	0.48	0.10	1.50	0	0	2.10	1.41	0.92	0.69	0.16
83	0.55	0.48	0.10	0.47	0.07	0	0	0	1.57	1.41	0.92	0.69	0.16
121	0.35	0.32	0.09	0.02	0.03	0	0	0	−0.11	0.10	0.07	0.06	0.03
122	0.34	0.32	0.09	0.02	0.02	0	0	0	−0.14	0.10	0.06	0.05	0.03
123	0.34	0.32	0.09	0.02	0.02	0	0	0	−0.17	0.09	0.06	0.05	0.02
124	0.34	0.32	0.09	0.02	0.02	0	0	0	−0.19	0.08	0.05	0.05	0.02
125	0.34	0.32	0.09	0.02	0.02	0	0	0	−0.21	0.07	0.04	0.04	0.02
													a

a = rooting depth.

SUMMARY

$$(T/T_M)_1 = \frac{0.060}{0.062} = 0.97; \quad (T/T_M)_2 = \frac{13.54}{13.65} = 0.96; \quad (T/T_M)_3 = \frac{6.22}{6.55} = 0.95$$

$$(T/T_M)_4 = \frac{4.77}{5.75} = 0.83; \quad (T/T_M)_5 = \frac{4.79}{14.83} = 0.32$$

$$Y/Y_M \text{ (dry matter)} = T/T_M \text{ (season)} = \frac{29.38}{41.36} = 0.71 \qquad \text{Equation [5]}$$

$$Y/Y_M \text{ (grain)} = (0.97)^{0.0} \cdot (0.96)^{0.4} \cdot (0.95)^{0.4} \cdot (0.83)^{0.4} \cdot (0.32)^{0.0} = 0.89 \qquad \text{Equation [6]}$$

Since evaporation of water from the soil is a process very different from transpiration, a different equation is used to estimate the relation of E to E_M

$$E = E_M/t^{\frac{1}{2}} \qquad\qquad [9]$$

where t is the time in days since the soil was last wet. All water for evaporation is used from the top soil layer and is subject to the restriction that the soil water cannot be decreased below the air dry water content.

Once the input (starting) information is known the estimates of T and E are made on a daily basis from the equations discussed above as shown in Table 4. To make these computations it is necessary to keep track of the water status of each soil layer. Drainage out of the profile, as happened on day 6 of the example shown in Table 4, occurs when the water added is greater than "AW − SWC" summed over all layers. The model, to simplify the situation, assumes that all the transpiration for any day comes from the one or two soil layers where the ratio of SWC/AW is largest. Thus the computations are not accurate on a day-to-day basis but are reasonably correct on a weekly basis. The value of E was the same as E_M on the days rain or irrigation was applied, otherwise E was less than E_M depending on the time since the last wetting. As the season progressed the ratio of T_M/E_M increased to a constant value of about 5 at midseason. For the example shown, T was near T_M for the first three growth stages and decreased to a low value in the last growth stage. This had more of an influence on predicted dry matter yield than on grain yield. Note that SWC did take on negative values. This is an indication that evaporation from the upper soil layer decreased the water content below wilting.

The data of Table 4 also show the effect of the active growing root. For this crop it was assumed that roots were growing into the soil to reach a stable depth at 90 days. Root extraction occurred in the second layer after the 24th day and in the third layer after the 46th day. The FORTRAN computer code for this model is shown in the appendix.

This model has been compared with many situations where measured data are available. Figure 5 shows a good agreement between measured and computed grain and dry matter yields for corn as reported by Hillel and Guron (1973) in Israel (Hanks, 1974). Hanks (1974) also showed good agreement between computed and measured grain yields in Nebraska (Fishbach and Somerhalder, 1972) where summer rainfall was considerable. Stewart *et al.* (1977) show a comparison of measured and computed yields, Figures 6 and 7, for Arizona, California, Colorado and Utah for two years. The agreement was very good for dry matter yields and less satisfactory for grain yields.

This model is very easy to use for testing the predicted yield as related to

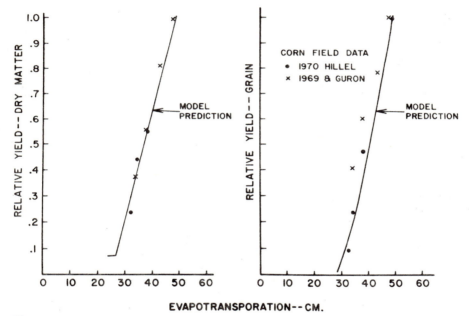

Figure 5. Comparison of measured and computed relative yield of dry matter and grain for the data of Hillel and Guron (1973) in Israel. Computations by the method of Hanks (1974).

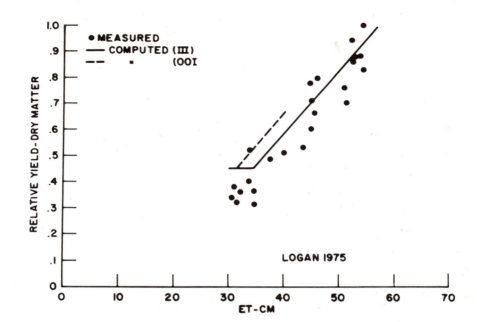

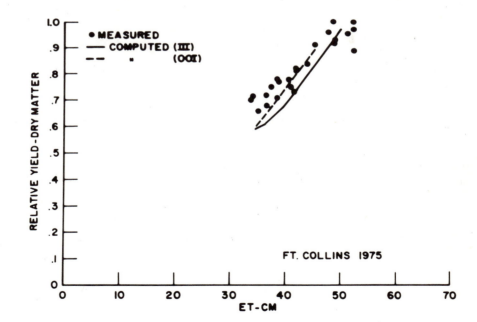

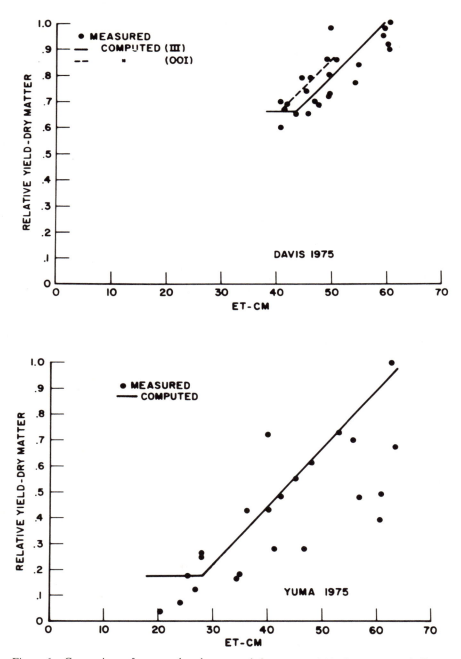

Figure 6. Comparison of measured and computed dry matter yield of corn grown in Logan, Utah; Davis, California; Ft. Collins, Colorado; and Yuma, Arizona in 1975. Data reported in Stewart *et al.* (1977). Computation H-1 refers to equation [5] of the method of Hanks (1974). Computations for III and OOI treatments were slightly different because of different irrigation applications. Treatments are described in Figure 1.

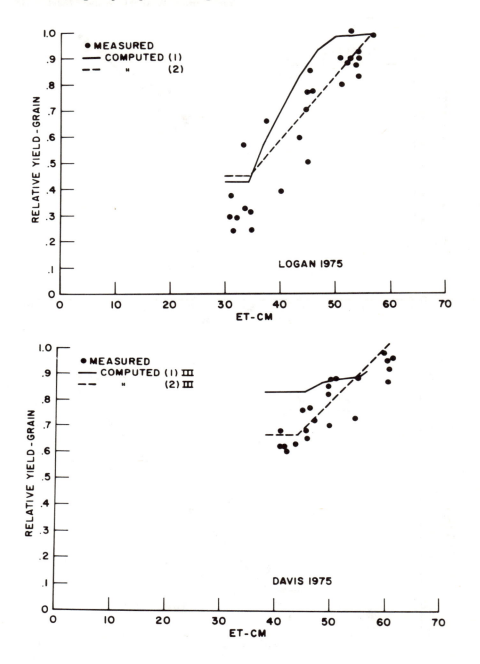

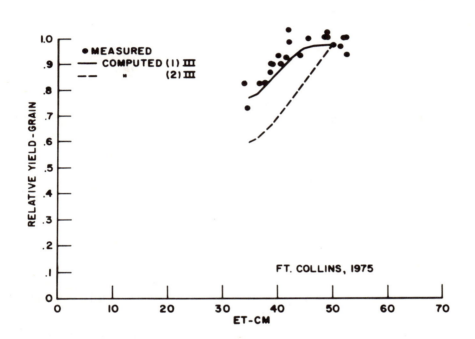

Figure 7.   Comparison of measured and computed grain yields of corn grown in Logan, Utah; Davis, California; and Ft. Collins, Colorado in 1975. Data reported in Stewart *et al.* (1977). Computations made by the method of Hanks (1974) using equation [5] for H-1 and equation [6] for H-2. Treatments are described in Figure 1.

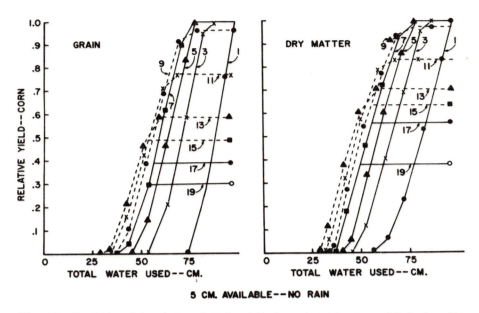

Figure 8. Simulation of the relation of relative yield of corn to total water used (irrigation plus soil water storage) as related to the number of days between irrigations (Hanks, 1974). Equation [5] used for dry matter and equation [6] for grain.

one of many possible input variables. This attribute is an important advantage in any model. It allows a simulated experiment to be performed at very little cost by changing only one variable at a time. Predictions of the influence of many management factors are thus possible. This is illustrated by Hanks (1974) for the irrigation interval in Figure 8. For corn where irrigation was applied frequently, relative yield was lower for a given water use (ET + drainage) than where irrigation was applied less frequently (up to 9 days). This difference was caused by more water being lost through evaporation where irrigation was frequent. Water lost by evaporation does not contribute to yield. Figure 8 also illustrates the problem that might occur if water were applied at very long intervals. In this situation the soil water storage is not enough to supply plant needs between irrigations so T/T_M is limited and, if the same amount of water is used, much of it is lost to drainage.

Hill *et al.* (1974) have used the model of Hanks (1974) and modified it to include a climatic prediction (growing degree days) of growth stages. They then tested for the best time of the year to apply a limited amount of water, assuming a given rainfall climatic situation like that of Mead, Nebraska in 1972. This is illustrated in Table 5 for two different planting dates. They also

Table 5. Predicted influence of incremental application levels of irrigation water on corn yield (Mead, Nebraska, 1972 data)

Planting date	Dates of application of 1 inch of irrigation (in order of selection)	Predicted yield (with irrigation on previously selected dates) percent of potential
May 4	None	33
	July 17	60
	July 19	74
	July 3	85
	July 7	93
	May 26	96
May 18	None	63
	July 23	85
	June 15	92
	June 15	96

predicted the influence of planting date on relative yield for the same situation and found that predicted grain yields were much more sensitive to planting date than were dry matter yields.

The model of Hanks (1974) was modified (Rasmussen and Hanks, 1978) to predict spring grain yields as influenced by irrigation. The modification included a growth stage prediction similar to that of Hill *et al.* (1974) as well as an adjustment for depth of planting. Several wheat varieties were grown under variable irrigation treatments and measured yields were compared with predicted yields. A further modification involved use of the following equation to estimate T_M:

$$T_M = a E_0 \qquad [10]$$

From planting to emergence, $a = 0$. From emergence to heading the value of a was varied from 0 to 0.9 linearly. From heading to soft milk stage a was assumed to be 0.9, after which it decreased linearly to zero at maturity.

To estimate E_M, the following equation was used:

$$E_M = b (E_0 - T_M) \qquad [11]$$

where $b = 1$ from planting until soft milk stage, after which time b decreased linearly to 0.2 at harvest (to account for shading).

The λ values were determined for a calibration set of data. The value of 0.25 for all growth stages gave the best fit for many combinations tried. This model was then used to predict yields for measured data for previous years and at a dryland site. Figure 9 shows good agreement between predicted and measured yields for both the calibration and validation data. The calibration data agree best, as would be expected. This comparison assumes that the yields measured in 1975 for no water limitations were the Y_M yields. The highest yields in 1972 were considerably higher than in 1975. Since the model is limited to yield effects due to water deficit, Y_M values should be found for each year. This was not done by Rasmussen and Hanks (1978) because it was not possible to predict future Y_M values. This points out an important aspect of this type of model — a reliable value for Y_M must be known. Some other method could possibly be used to make this prediction.

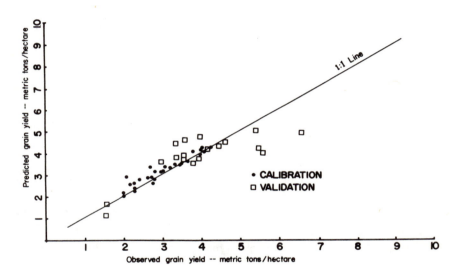

Figure 9.   Comparison of observed and predicted spring wheat yields as predicted by the model of Hanks (1974) using equation [6] as reported in Rasmussen and Hanks (1978).

Hill, Johnson and Ryan (1979) developed a model for estimating soybean yield using a similar approach to that of Hill *et al.* (1974) which was a modification of the Hanks (1974) model. A phenology clock for soybeans which predicted plant growth stage progress was linked to yield prediction equations. Yield was predicted as a function of relative transpiration during each of five growth periods: planting to emergence, emergence to beginning flowering, beginning flowering to beginning podfill, beginning podfill to end of flowering, and end of flowering to maturity. The phenology clock was adapted from work in Missouri by Major *et al.* (1975) in which growth stage progress was related to temperature and day length.

Preliminary calibration attempts of the model indicated that equation [6] alone was not satisfactory for predicting yields with extremely late planting dates for which the relative transpiration was high but yield was very low. Hill *et al.* (1979) hypothesized that this was due to insufficient accumulation of dry matter. There may be a threshold value of accumulated transpiration below which a corresponding decrease in total dry matter production limits bean yields. They adapted equation [6] to the soybean yield model by using a seasonal yield factor (SYF) and a lodging factor (LF). The seasonal yield factor accounts for reductions in yield as a result of late planting. This accounts for the potential loss in yield from lodging caused by too much available water during the vegetative growth period.

Bean yield was thus estimated using the following modification of equation [6]:

$$Y/Y_M = (T/T_M)_1^{\lambda_1} \cdot (T/T_M)_2^{\lambda_2} \cdot (T/T_M)_3^{\lambda_3} \cdot (T/T_M)_4^{\lambda_4} \cdot$$

$$\cdot (T/T_M)_5^{\lambda_5} \cdot SYF \cdot LF \qquad [12]$$

The Jensen-Haise (Jensen, 1973) equation was used to calculate the reference crop evapotranspiration from which the values of T_M, E_M, and T and E were estimated as previously discussed. The combined soybean phenology-yield model was calibrated using data from the Missouri field tests (Johnson *et al.*, 1973a, 1973b). It was found in the calibration process that the values of the λs were dependent not only on growth stage but also on maturity group.

Parameter values which gave the best fit model are presented in Table 6. The potential yields shown were estimated values. The relationship between relative yields predicted by the model (equation [12]) and actual field test relative yields is shown in Figure 10. Careful examination of planting dates,

Table 6. Calibrated parameter values for equation [12] from Missouri soybean field tests (Hill *et al.*, 1979)

Maturity Group	Parameter values					
	equation [6]					Potential yield
	λ_1	λ_2	λ_3	λ_4	λ_5	Y_M (quintals/ha)
II	0	0	0.380	0.05	0.410	51.8
III	0	0	0.075	0.03	0.475	60.5
IV	0	0	0	0.30	0.400	63.9

sites and yields for each individual point did not reveal a consistent selective bias of the overall model to any given set of field considerations. In particular, the model predicted both the highest and the lowest field yields for Maturity Group II with comparable closeness of fit, whereas for Maturity Group IV the highest and lowest yields were in the worst fit category.

Data from four year and site combinations other than those used in calibration were obtained for testing, or verifying, the soybean phenology-yield model. The results of this verification presented in Table 7, in which the parameters were fixed at the values of Table 6, indicate good agreement of predicted and measured yields with r^2 values greater than 0.96. Especially pleasing was how well the model simulated the effect of irrigation observed at the McCredie, Missouri site.

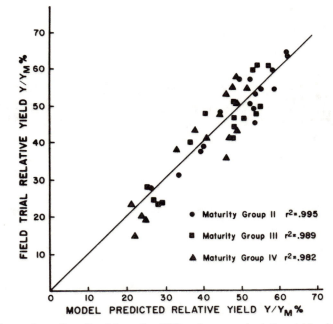

Figure 10. Comparison of predicted (equation [18]) and measured relative yields of soybeans by the model of Hill *et al.* (1979). Data were from several varieties grown in 1971 and 1972 at three different sites.

The model-predicted yields were higher than actual yields for the July 1 and 11 planting dates at Columbia and Spickard, Missouri. This may have been as a result of the assumption that the beginning soil moisture was at field capacity when it probably should have been lower for these later plantings.

Table 7. Soybean yields from model verification

Location	Year	Planting date	Soybean maturity group					
			II		III		IV	
			actual	model	actual	model	actual	model
			. . . . .yield (quintals/ha).					
Columbia,	1973	4/29	19.5	27.6	24.9	25.4	24.2	25.8
Missouri		5/15	28.2	24.8	26.9	24.6	23.5	22.7
		5/31	26.2	20.8	23.5	23.7	18.8	23.2
		6/21	20.8	24.4	23.5	27.0	22.9	22.4
		7/11	16.1	23.2	19.5	25.2	16.8	23.4
Spickard,	1973	5/11	30.3	33.7	31.6	31.0	24.9	32.5
Missouri		6/8	24.9	30.3	31.6	30.5	28.9	30.3
		6/25	27.6	26.3	24.2	26.5	25.6	23.4
		7/1	21.5	25.0	20.2	25.0	18.8	22.2
Johnston, Iowa	1976	5/4	29.4	27.3	29.2	24.6	23.3	23.9
McCredie,	1973	5/16[1]	—	—	—	—	19.5	19.1
Missouri		5/16[2]	—	—	—	—	27.6	27.5
Regression (r²)			0.963		0.985		0.973	

[1] Non-irrigated.
[2] Irrigated with 2.5 cm July 18, 2.5 cm August 24.

Kanemasu Model

Kanemasu and co-workers have also developed models of crop response to irrigation (Morgan *et al.*, 1979, Kanemasu *et al.*, 1976). These models combine water balance and evapotranspiration models (Kanemasu *et al.*, 1976) with a crop response model in a somewhat different way from those described above.

They computed potential transpiration, T_M, as

$$T_M = 1.74(1 - \tau)[S/(S + \gamma)][0.86 R_s - 103.9]/590 \quad \text{for}$$

$$\text{LAI} < 3.0 \qquad\qquad\qquad\qquad [13]$$

$$T_M = (\alpha - \tau)[S/(S + \gamma)][0.848 R_s - 144.5]/590 \quad \text{for}$$

$$\text{LAI} \geq 3.0 \qquad\qquad\qquad\qquad [14]$$

where α is a crop and location dependent variable (1.35, 1.28, 1.45 and 1.35 for corn, sorghum, soybeans, and winter wheat, respectively, in Kansas), S is the slope of the saturation vapor pressure curve, γ is the psychrometer constant and R_s is solar radiation. For equation [14] the coefficients are changed from 0.848 to 0.766 and from 144.5 to 99.9 if the growing degree days are greater than 1690. This approach is a modification of the method of Penman which has in turn been modified by Priestly and Taylor (1972) and by Jury and Tanner (1975). The fraction of energy supplied to the soil surface (τ) depends on shading and leaf area index (LAI) by

$$\tau = \exp(-0.389\,\text{LAI} + 0.15) \qquad [15]$$

During hot days, T_M is increased by an advective correction.

Evaporation directly from the soil is computed using two equations. During "constant rate", where evaporation losses are determined by climatic factors, the following equation is used:

$$E_M = \tau\left[S/(S + \gamma)\right] R_n/590 \qquad [16]$$

where R_n, net radiation, can be directly measured or approximated from measurements of R_s, as given in equations [13] and [14]. However, where water cannot be transported to the soil surface fast enough to meet climatic demands, another equation is required, as

$$E = \left[ct^{1/2} - c(t-1)^{1/2}\right] \qquad [17]$$

where "c" depends on the soil's water transmitting properties. In practice the model computer program uses equation [17] unless irrigation or rain exceeds 0.6 mm, when it switches to equation [16]. Equation [16] is used each day until the sum since the last rain or irrigation reaches a threshold value u (which is mainly dependent on soil texture). After this time equation [17] is used. The values of u, c and other soil properties are given in Table 8 for several Kansas soils.

To account for the limitation of transpiration as water content decreases, an equation like that of Hanks (1974) (equation [8]) is used, except that b is 0.35 rather than 0.5 and only one soil layer is assumed. Thus a daily computation of the water balance is made by using a procedure similar to that described earlier for the Hanks model. There is, however, a correction made for effective precipitation, to account for runoff.

Thus, an estimate of the soil water status can be made from knowledge of the soil variables given in Table 8, climatic variables (T_{max}, T_{min}, R_s and P)

Table 8. Approximate values of u, c, field capacity (FC) and maximum available water content (AW) in the root zone (150 cm) for several Kansas soils

Soil type	u	c	FC	AW
	cm	cm day$^{-1/2}$	cm	cm
Carurle sand	0.50	0.168	8.2	3.7
Florence stony loam	0.58	0.289	52.1	17.9
Manter fine sandy loam	0.90	0.241	23.0	11.5
Mansic silty clay loam	0.92	0.353	33.4	17.6
Muir silty clay loam	1.02	0.327	54.6	25.7
Harney silt loam[1]	1.12	0.336	46.8	20.7
Harney silt loam[2]	1.26	0.353	56.0	32.0
Lancaster clay loam	1.65	0.373	54.4	28.1

[1] Hays, Kansas.
[2] Minneola, Kansas.

and crop variables (LAI, rooting depth, time of planting, etc.). This information is then used to compute crop growth by a separate function.

The general growth response equation for corn is

$$Y = \prod_{t=1}^{t_M} \Gamma(t)^{\sigma_t} \, Y_{Mt} \qquad [18]$$

where t is time in days, Y_M represents potential yield at maturity t_M, $\Gamma(t)$ represents relative growth in potential yield in time period t when water is not limiting, and σ_t is a function relating growth (relative to potential growth) to the level of available soil moisture.

In this model the plant growth process is divided into vegetative and reproductive growth phases. The vegetative phase corresponds to growth stages 0 to 4.5 and the reproductive stage corresponds to 4.5 to 10 on Hanway's (1963) scale. The relative dry matter accumulation during the vegetative stage for corn is in the form

$$(Y/Y_M)_V = \exp(-1.7 + 0.094t) \qquad [19]$$

After stage 4.5 is reached, the relative dry matter accumulation in the reproductive stage is determined by another equation of the form

$$(Y/Y_M)_R = \exp(-3.573 + 0.109\,(t_R - t_R^2/2t_{RM})) \qquad [20]$$

where t_R is the time in days into the reproductive stage, and t_{RM} is the total time in the reproductive stage. Figure 11 shows the comparison of model simulations and Hanway's data for both stages.

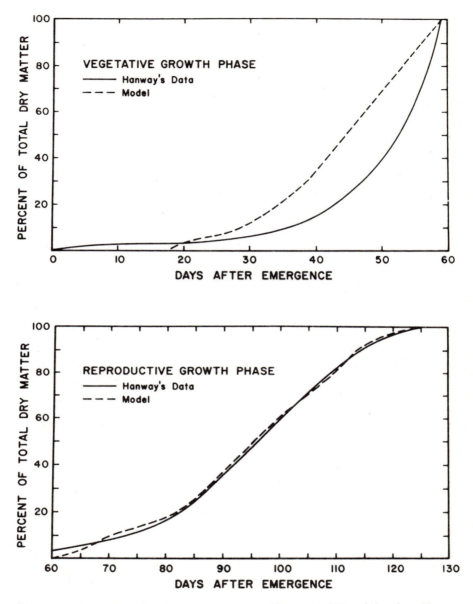

Figure 11.   Comparison of model predictions of Morgan *et al.* (1979) and data from Hanway (1963) for corn for vegetative and reproductive growth stages.

The coefficient σ_t is a crop response function related to available soil water. The best fit function is shown in Figure 12, which starts at 1.0 for soil water at field capacity and stays at about 0.8 until the relative soil water falls below about 0.3

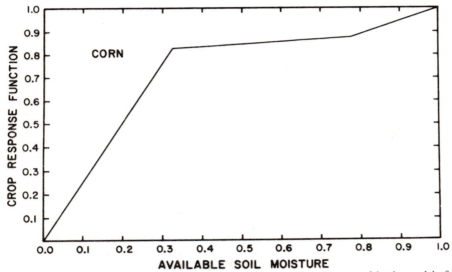

Figure 12. Relation of crop response function to available soil moisture used in the model of Morgan *et al.* (1979).

A comparison of predicted and actual grain yields for several years and locations in Kansas is shown in Figure 13 which indicates a reasonably good relation. This model has also been used to predict the result of irrigation frequency and amount of yield. These results are shown in Figure 14 and Table 9.

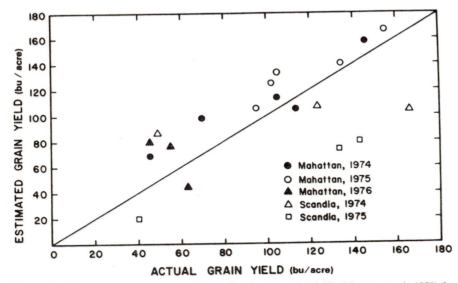

Figure 13. Comparison of measured and predicted corn grain yields (Morgan *et al.*, 1979) for several years and locations in Kansas.

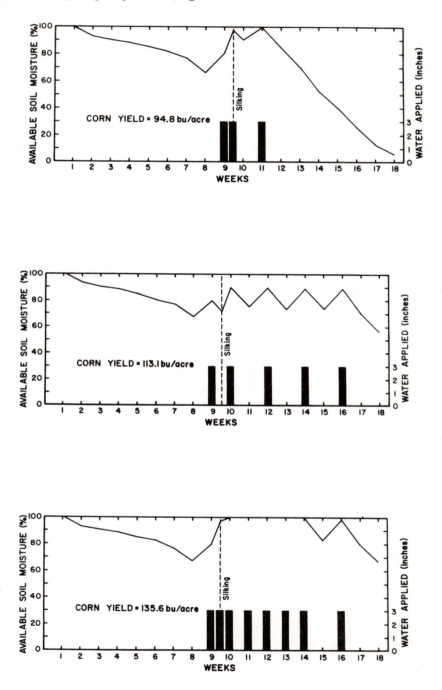

Figure 14.   Simulation of the available soil moisture as related to time for several irrigation schedules as predicted by the method of Morgan *et al.* (1979). Yields and irrigation water applied are also shown.

Table 9. Results of simulated irrigation schedules based on estimated corn response model of Morgan *et al.* (1979)

Simulated run number	Water applied to soil in root zone, cm	Water pumped, cm	Reduction of available moisture in root zone over growing season	Total water used, cm	Predicted yield, kg/ha	Variable cost of irrigation, $/ha	Predicted net return, $/ha*
1	30.5	50.8	21.1	71.9	7400	78.23	92.28
2	38.1	63.5	15.1	78.6	7865	97.79	109.58
3	45.7	76.2	11.8	88.0	8034	117.35	94.35
4	53.3	88.9	11.8	100.7	8223	136.91	89.06
5	61.0	101.6	11.2	112.8	8511	156.46	102.88
6	38.1	63.5	13.3	76.8	7099	97.79	48.62
7	30.5	50.8	18.1	68.9	5699	78.23	25.01
8	30.5	50.8	18.0	68.8	6026	78.23	− 7.34
9	30.5	50.8	19.2	70.0	7136	78.23	71.35
10	30.5	63.5	10.5	74.0	5739	97.79	− 60.39
11	38.1	63.5	11.4	74.9	6379	97.79	− 9.38
12	45.7	76.2	8.1	84.3	7930	117.35	95.24

* Not including land charge.

Transpiration Model—Soil Water Flow

The models discussed up to this point use an extremely simplified approach to estimating soil water influences on plant uptake and water flow. They are physically limited because they do not allow water flow to be influenced by time, nor do they allow upward flow. They also assume that root uptake is related to water availability which is a simple function of soil water content. These simplifications will not allow for the influence of a water table near the surface nor for salinity to be included. Thus, a more fundamental method is needed in some instances.

Hanks and co-workers (Nimah and Hanks, 1973; Childs and Hanks, 1975) have described such a model for one-dimensional soil water flow, as follows:

$$\frac{\partial \theta}{\partial t} = \frac{\partial}{\partial z} K(\theta) \frac{\partial H}{\partial z} + A(z) \tag{21}$$

where $K(\theta)$ is hydraulic conductivity (a function of water content), θ is volumetric water content (expressed as a fraction), t is time, z is soil depth, H is hydraulic head (the sum of matric and gravitational head) and $A(z)$ is a root extraction term. The root extraction term is given as

$$A(z) = \frac{[H_{Root} + (1.05 \cdot z) - h(z, t) - s(z, t)] \cdot RDF(z) \cdot K(\theta)}{\Delta x \cdot \Delta z} \tag{22}$$

where H_{Root} is the effective root water head at the soil surface, the term "$1.05 \cdot z$" is a correction to the root water head at other soil depths, h is the matric head, s is the osmotic head which accounts for salinity, RDF is the root density function, Δx is the distance from the root surface to the point in the soil where h(z, t) and s(z, t) are measured (arbitrarily assumed to be 1.0 cm) and Δz is the depth increment.

The root extraction term is a dynamic term influenced by soil, plant and climatic conditions. A numerical method was devised to solve equations [21] and [22] on a high speed digital computer. In this method the value of H_{Root} is "hunted" for until the integrated plant root extraction over the soil profile is equal to T_M, provided that the value of H_{Root} is greater than some lower limiting value at which "wilting" occurs. When the soil water is low or the salinity is high, then T will be less than T_M and yield would be decreased. Thus in the model, H_{Root} is bounded on the wet end by $H_{Root} = 0.0$ and on the dry end by $H_{Root} = H_{Wilt}$. We have assumed a value for H_{Wilt} of $-12,000$ to $-20,000$ cm.

This model used the same scheme as Hanks (1974) to determine E_M and T_M with the modification that the average daily values are converted (internally in the computer program) to zero during the 12 night hours and vary like the upper half of a sine wave during the 12 day hours. Thus $T = T_M$ during the night regardless of soil conditions because $T_M = 0$. In the daylight hours $T = T_M$ for some time depending on the complex interaction of soil $(K(\theta), \theta, h)$, plant (H_{Root}, RDF) and climatic factors (T_M, E_M). This model computes on a much smaller time scale (0.024 to 2.4 hours) compared to the one day of the models described earlier. It is much more realistic but also costs more to run.

The relation of E to E_M is also dynamic and much more complicated than for the models described earlier. If the surface water content and associated matric head is greater than for air dry conditions (h of about -10^{-6} cm), then $E = E_M$. When water cannot be transported to the surface fast enough to keep the surface water content above θ (air dry), then $E < E_M$.

When irrigation or rain occurs, it is not only necessary to know the amount (like Hanks, 1974), but also to know the intensity (unlike Hanks, 1974). The water content of the soil surface increases, when water is applied, to some value which cannot exceed the saturated water content and associated matric head (h = 0). If water application is relatively low, the surface water content will not reach saturation but will tend to a water content where $K(\theta)$ = application rate. However, if the application rate is high (higher than $K(\theta)$ saturated) the soil at the surface will wet up to the saturated water content and will not be able to transmit all the water applied to the soil. Thereafter ponding and runoff will occur.

The root extraction term includes the salinity effect if the main salinity effect is osmotic only. Thus no specific ion or other effects are considered. It is also necessary that the associated flow of salt be accounted for. This is done by solving the following salt flow equation:

$$\frac{\partial(\theta C)}{\partial t} = \frac{\partial}{\partial z}\left[D(\theta, q)\frac{\partial C}{\partial z} + qC \right] \tag{23}$$

where C is salt concentration, $D(\theta, q)$ is a combined diffusion and dispersion coefficient, and q is volumetric water flux (computed from equations [22] and [23] as discussed in more detail by Childs and Hanks (1975). No precipitation of salt or dissolution is considered in this treatment.

The model assumes no salt flow into the plant, so salt can be concentrated by root extraction of pure water. Salt concentration is influenced by the concentration of the irrigation water applied and water flow in or out of the bottom of the root zone. Nimah and Hanks (1973) give an example of the data required $(K(\theta)$ vs θ and h vs $\theta)$ for a typical soil.

The same approximations regarding relative yield as related to relative transpiration are used in this model as those discussed for the Hanks (1974) model. The difference is in how the transpiration, evaporation and interactions are handled.

The input data requirements to solve a problem using this model can be grouped as basic properties of soils and plants, initial conditions and boundary conditions. The specific data needed are as follows:

1. *Soil and plant properties*—data of $K(\theta)$ vs θ and h vs θ are required (soil properties), as well as data required to compute $A(z)$ (equation [22]). If roots grow during the season, RDF(z) must be changed with time as well as depth. As discussed by Nimah and Hanks (1973), RDF(z) is the fractional amount of active roots in a particular depth increment. Also needed are the values of H_{Wilt} and the air dry matric head.

2. *Initial conditions*—the value of any water content and soil solution concentration vs depth at the beginning must be known.

3. *Boundary conditions*—the potential climatic conditions at the soil plant surface must be known for the entire time of the simulation. E_M and T_M vs time and irrigation and rainfall amounts and time increments similar to the Hanks (1974) model are required. It is also necessary to know something about the lower boundary. If there is a water table at the lower boundary, h = 0 at that depth provided it does not change with time. With this boundary condition, upward flow or drainage could occur. Another boundary condition provided for is an impervious layer where no flow occurs. It is also possible to make the

soil deep enough so that an impervious layer at the bottom will provide enough soil water storage above the layer, thus allowing no change.

This model has been tested under various conditions by Nimah and Hanks (1973), Childs and Hanks (1975), and Wolf (1977). Figure 15 shows that computed soil water agrees well with measured soil water over a yearly irrigation situation with alfalfa (Nimah and Hanks, 1973). Wolf (1977) used the model to see if it could reasonably estimate the results of the irrigation timing and salinity study reported by Stewart *et al.* (1977). Figure 16 shows a comparison between computed and measured results where irrigation amount

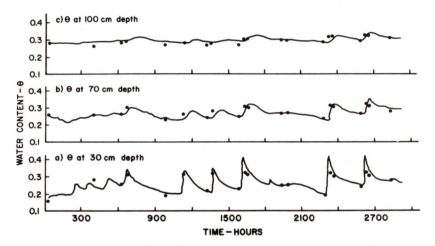

Figure 15. Comparison of predicted (lines) and measured soil water content for a season at three depths using the model of Nimah and Hanks (1973).

and timing were varied. Excellent agreement between computed and measured data was found where soil salinity was not high. However, the model overestimated yields where salinity levels were high. Field observations indicated that the salinity influence early in the year was greater than computed by the model. The model assumed the influence of salinity to be osmotic only and assumed the same effect throughout the year.

Many simulations were made by Childs and Hanks (1975) on the effect of several irrigation, salinity, and root distribution variables on relative dry matter yield. Figure 17 shows the influence of root depth, original soil salinity, and irrigation and rain on relative yield. The data were for the field situation of Nimah and Hanks (1973) where there was a possibility of water flow up

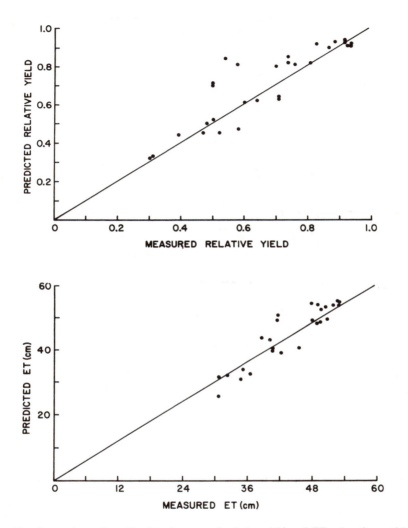

Figure 16.   Comparison of predicted and measured relative yield and ET using the model of Childs and Hanks (1975) when treatments S_4 and S_5 were salinized and treatments S_0 through S_3 were relatively nonsalinized as reported by Wolf (1977). Data are given in Stewart et al. (1977) and treatments described in Figure 1.

from a water table at about 2 m. Because of upward water flow, the influence of rooting depth on relative yield was large where the amount of water applied was much less than water lost through evapotranspiration. When the amount of irrigation was higher, the ET of all treatments gave the same relative yield of 1.0. Figure 18 shows the predictions made of drainage or upward flow for the same data situation as Figure 17. The deep rooted crop shows as much as

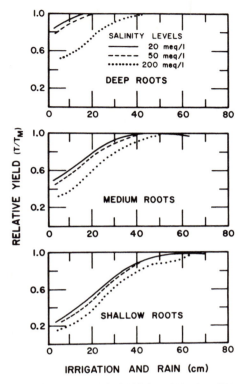

Figure 17.   Simulations made by the model of Childs and Hanks (1975) of the influence of rooting depth, initial soil salinity and amount of irrigation and rain on relative dry matter yield.

four times the upward flow as the shallow rooted crop. The relative effect of initial salinity on upward flow is greater for the deep roots than for shallow roots. The shallow rooted crop always shows a lower yield than deeper roots because less of the stored soil moisture is available due to less rooting depth.

One of the significant advantages of the model is the possibility of simulating situations which are impractical to measure. This is illustrated in Figure 19 (Childs and Hanks, 1975), which shows the influence of two levels of irrigation on relative yield over several years. The same climatic data were assumed for all years. At the highest irrigation level (43 cm/year, which was about 5 cm less than ET) the relative yield did not start to fall off until after about 6 years in spite of continuous salt buildup during the period. Figure 19 shows that the lower irrigation rate gave an almost linear decrease in yield and a corresponding rise in salinity over 4 years, with a leveling off thereafter. With the same amount of water applied, upward flow decreased with time to zero. If the simulations had been continued, there would have been some drainage

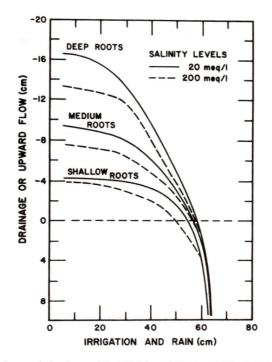

Figure 18.   Simulation made by the model of Childs and Hanks (1975) of the influence of rooting depth, initial soil salinity, and the amount of irrigation and rain on relative yield.

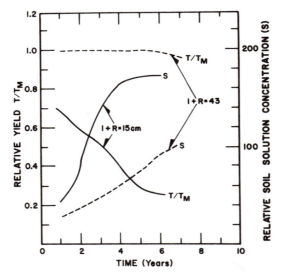

Figure 19.   Simulation of the effect of time and the amount of irrigation and rain on relative yield (Childs and Hanks, 1975).

until salt balance occurred. Thus, an equilibrium situation would have been achieved. This assumes uniform climatic conditions which may be somewhat unrealistic.

Also of interest is a simulation done by Childs and Hanks (1975) involving irrigation uniformity. Under any practical irrigation system, water is applied unevenly. While the average application rate may indicate no salt buildup because drainage occurs, there may be areas receiving little enough water so that no drainage occurs and thus there is a salt buildup. To illustrate this, two situations were considered. The first was a solid set sprinkler irrigation system with about as high uniformity as presently possible with a coefficient of uniformity (CU) of about 0.88. The coefficient of uniformity is defined as CU = 1 − DV/IA where IA is the average irrigation rate and DV is the average deviation from IA. It was assumed that these sprinkler lines were left in the same place year after year. The second situation was a very poor gravity system similar to the one used in the on-farm study of Nimah and Hanks (1973) before the sprinklers were installed. It was estimated that this system had a CU = 0.42. Table 10 shows the calculations of relative yield, salt outflow and final salinity at the end of 1 and 5 years for both situations tested for one

Table 10. Example of uniformity predictions — 5-year sequence for shallow rooted crop

Irrigation and rain, cm	Area,* %	Year 1			Year 5		
		Rel. yield	Salt outflow, kg/ha	Final soil solution salinity, meq/liter	Rel. yield	Salt outflow, kg/ha	Final soil solution salinity, meq/liter
. . . . CU = 0.88 Parabolic Distribution							
40.4	10.4	0.85	—	27.6	0.84	—	70.8
46.7	24.8	0.92	—	25.7	0.91	—	52.5
53.1	29.6	0.96	—	23.9	0.96	—	38.7
59.5	24.8	0.99	516	22.4	0.99	67	29.6
65.8	10.4	1.00	1233	21.2	0.99	1434	24.0
Average		0.95	247		0.95	314	
. . . . CU = 0.42 Rectangular Distribution							
10.6	20.0	0.50	—	32.3	0.39	—	161.6
31.9	20.0	0.75	—	29.9	0.68	—	50.5
53.1	20.0	0.96	—	23.9	0.96	—	38.7
74.3	20.0	0.96	2556	20.4	0.96	2666	21.1
95.6	20.0	1.00	8138	20.0	1.00	8646	20.0
Average		0.83	2175		0.80	2285	

* The percent of the area having the average irrigation and rain in the column to the left.

irrigation rate. For simplification of illustration, the area is broken down into fifths. The simulations show that the yield varies because of nonuniform water application and that the high uniformity gave higher yields than the low uniformity system. After 5 years there was a larger decrease in yields, on the average, for the lower uniformity system, caused by the higher salt buildup. In addition, more salt outflow into the drains occurred for the low uniformity than the high uniformity system.

The multiyear calculations illustrate how model simulation can achieve answers to questions that it would be impractical to attempt to solve in the field for economic reasons. These simulations have also been used as a basis for economic studies by Hanks and Andersen (1978) in which they have computed costs of limiting salt outflow·in drainage water.

Childs *et al.* (1977) have developed another model for corn which can be used to predict seasonal yield from irrigation input variables and which is much more sophisticated with regard to plant growth relations than the models described above. The soil water flow part of their model is a modification of that reported by Childs and Hanks (1975) and Nimah and Hanks (1973). They also use the approach of separating evapotranspiration into its components. Evaporation (potential), E_M, from the soil for a corn crop is computed as

$$E_M = ET_{MA} \cdot \exp (0.438 \cdot LAI) \qquad [24]$$

where ET_{MA} is potential evapotranspiration using alfalfa as a base, and LAI is leaf area index. Actual transpiration was computed in the same way as reported by Nimah and Hanks (1973).

Potential transpiration, T_M, is calculated using the following equation:

$$T_M = ET_{MA} \cdot (1 - E_M) \cdot K_{CT} \qquad [25]$$

where K_{CT} is a crop coefficient related only to transpiration. Actual transpiration was computed from the following general equation:

$$T = \frac{\psi_1 - \psi_s}{r_s + r_p} \qquad [26]$$

where ψ_1 is the water potential in the leaf, ψ_s is the soil water potential (sum of osmotic and matric potentials), r_p is the plant flow resistance, and r_s is the soil flow resistance (the inverse of the hydraulic conductivity). The above equation cannot be solved as it is because ψ_1 is not known, and ψ_s and r_s vary with depth. To solve equation [26], the following working equation is used:

$$T = (\psi_{1min} - H_{Root})/(r_p \cdot LAI)$$

(subject to the restriction $T \leq T_M$)

where ψ_{1min} is the minimum allowable value for leaf water potential, H_{Root} was calculated using the method of Childs and Hanks (1975) assuming $T = T_M$ and r_p was calculated as the product of the plant flow resistance per unit path length (444 cm hr/cm^2 assumed for corn) and the average path length of water flow (150 cm assumed for corn). Note that in the above equation the restriction is imposed that $T \leq T_M$. If T is computed to be greater than T_M using the above equation, T is taken as equal to T_M and the equation is rearranged to allow computation of ψ_1 (instead of ψ_{1min}). Thus, if $T = T_M$, the value of ψ_1 must be greater than ψ_{1min}. Thus, the above approach differs from that of Nimah and Hanks (1973) in that a limiting value of leaf water potential is assumed (-13 bars for corn) to cause $T < T_M$, rather than a limiting value of H_{Root} as used by Nimah and Hanks (1973). The above approach also produces a value for leaf water potential which is used to compute dry matter production by the following equation:

$$Y_{Dry\ Matter} = PH - RS - AC \qquad [27]$$

where PH is photosynthesis, AC is available carbohydrates in the leaf and RS is respiration.

Photosynthesis is taken as being equal to a complicated function of radiation, sunlit leaf area, available carbohydrate, and leaf water potential (Childs *et al.*, 1977). Photosynthesis is the only component in the above equation related to soil water (and thus irrigation) through the relationship to leaf water potential. Respiration is computed as a complicated function of temperature, above ground dry matter and available carbohydrates in the leaf.

To compute field yields for comparison with measured data, equation [27] is solved on a daily or more frequent time scale. Above ground crop growth is assumed to equal half of dry matter production. Before tasseling this amount is divided equally between leaves and stems. After tasseling all the above ground dry matter is assumed to go to ear growth. Grain yield is then calculated from ear growth using a cob to grain ratio of 0.18.

Plant growth stages are assumed to be determined from degree day data. Degree days are accumulated when air temperatures are lower than 35°C and leaf water potential is above -13 bars. Fifty degree days after planting are allowed for seed germination and emergence. Tasseling was assumed when the degree day accumulation equaled 835 and maturity was assumed at 1545 degree days.

Comparison of measured and predicted corn yields were made by Childs *et al.* (1977) for an irrigation study done at Mead, Nebraska, as shown in Table 11. The comparison between measured and predicted yields is excellent. The

Table 11. Predicted (Childs *et al.*, 1977) and measured grain yields and plant growth for five irrigated treatments for Mead, Nebraska, 1975

Treatment	Leaf area		Stover weight (dry weight)		Ear weight (dry weight)		Grain yield (15.5% moisture)	
irrigation applied, cm/day	actual, dm²	predicted, dm²	actual, g	predicted, g	actual, g	predicted, g	actual, kg/ha	predicted, kg/ha
0.76	72.5	75.6	109.0	104.0	163.0	158.0	7500	7480
0.61	76.7	75.6	108.0	104.0	141.0	158.0	7260	7480
0.38	75.2	75.6	98.0	104.0	160.0	157.0	7050	7410
0.25	74.4	75.6	114.0	104.0	113.0	120.0	5230	5680
0.00	73.4	75.5	98.0	104.0	54.0	55.0	2490	2620

Variety: Pioneer 3366

influence of the irrigation treatment is clearly shown. Comparison with irrigation data at other locations and in other years were made with acceptable agreement between measured and predicted yields. This poorer agreement may be partially due to inadequate field data required by the model which then had to be guessed at. Nevertheless, it appears that the approach is a promising way of predicting detailed yield components as influenced by irrigation. Modifications for other crops and situations are needed.

Appendix

FORTRAN computer coding for the model of Hanks (1974). Also included are sample card input data. The data on the cards are as follows:

Card 1   DAYS, AIRDRY, AWFAC, RTDAMX, POTPRO, VARY
Card 2   N, NN, IR, IET
Card 3   E array (5 values)
Card 4   BGSM array (5 values)
Card 6   THK array (5 values)
Card 7   WHC array (5 values)
Card 8–9   Rain array (28 values)
Card 10–12 ET array (54 values)
Card 13–14 GIRR array (21 values)

```
      IF(WADD.LE.SEV) GO TO 56
      FRQI=DA
      GO TO 15
  56  SEV=WADD
  15  DO 80 I=1,5
      BGSM(I)=BGSM(I)+WADD
      IF(BGSM(I).LT.THK(I)*WHC(I)) GO TO 3
      WADD=BGSM(I)-THK(I)*WHC(I)
      BGSM(I)=THK(I)*WHC(I)
      IF(I.LT.5) GO TO 80
      CDRAIN=CDRAIN+WADD
  80  CONTINUE
   3  IF(BGSM(1)-SEV.LT.AIRDRY) GO TO 83
      BGSM(1)=BGSM(1)-SEV
      GO TO 74
  83  SEV=BGSM(1)-AIRDRY
      BGSM(1)=AIRDRY
  74  DO 90 I=1,5
  90  WC(I)=BGSM(I)/(THK(I)*WHC(I))
      KI=1
      TRAN=EVAP-SEVN
      DEPROT=DEPMAX/(1.+EXP(6.-12.*DA/RTDAMX))
  77  THIK=0.
      IK=1
      DO 81 I=1,5
      IF(I.EQ.1) GO TO 84
      IF(WC(I).GT.WC(I-1)) IK=I
  84  THIK=THIK+THK(I)
      IF(THIK.GE.DEPROT)GO TO 82
  81  CONTINUE
  82  IF(BGSM(IK).LE.0.) GO TO 9
      IF(WC(IK).GT.AWFAC) GO TO 75
      TRN=TRAN*WC(IK)/AWFAC
      IF(BGSM(IK)-TRN.LT.0.0) GO TO 76
      BGSM(IK)=BGSM(IK)-TRN
  78  CEVAP=CEVAP+TRN
      IF (KI.GT.1) GO TO 9
      KI=KI+1
      TRAN=TRAN-TRN
      WC(IK)=BGSM(IK)/(THK(IK)*WHC(IK))
      GO TO 77
  75  BGSM(IK)=BGSM(IK)-TRAN
      CEVAP=CEVAP+TRAN
      GO TO 9
  76  TRN=BGSM(IK)
      BGSM(IK)=0.0
      GO TO 78
   9  CET=CET+EVAP-SEVN
      CSEVP=CSEVP+SEV
      PI=DA/DAYS
      IF(PI.LT.DD) GO TO 26
      CPROD=CPROD*((CEVAP-C1)/CET)**E(J)
      C(J+5)=((CEVAP-C1)/CET)
      CPD=CPD+CEVAP-C1
      CFD=CFD+CET
      C(J)=CPROD
      DEPL=0.
      ETMAX=CET+CSEVP-C2
      ETDEF=CET-CEVAP+C1
      ETACT=CSEVP+CEVAP-C1-C2
      WATADD=CWADD-C4
```

```
      C4=CWADD
      C2=CSEVP
      DO 102 I=1,5
  102 DEPL=BGSAV(I)-BGSM(I)+DEPL
      DEP=DEPL-C3
      C3=DEPL
      IF(NN.NE.0)WRITE(6,101)
      WRITE(6,32)DA,EVAP,CEVAP,CSEVP,CIRR,RAIN,CDRAIN,CWADD,WC(1),WC(2),
     1WC(3),WC(4),WC(5),ETMAX,ETDEF,DEP,ETACT,WATADD
      C1=CEVAP
      IF(DA.GE.DAYS) GO TO 26
      J=J+1
      DD=DD+DDST(J)
      CET=0.0
   26 CKC=(CEVAP-CK1+CSEVP-CK2)/EVAP
      CK1=CEVAP
      CK2=CSEVP
      IF(NN.NE.0)WRITE(6,32)DA,EVAP,CEVAP,CSEVP,CIRR,RAIN,CDRAIN,CWADD
     1,BGSM(1),BGSM(2),BGSM(3),BGSM(4),BGSM(5),CKC
      DA=DA+1.
      IF (DA.LT.ET(KE)) GO TO 52
      SEVN=ET(KE+2)
      EVAP=ET(KE+1)
      KE=KE+3
      IF(SEV.GT.SEVN)SEV=SEVN
   52 SEV=SEVN/(SQRT(DA-FRQI+1.))
      IF(DA.LE.DAYS) GO TO 2
      CPD=CPD/CFD
      DA=DA-1.0
      CWADD=CSEVP+CEVAP+CDRAIN
      WRITE(6,30)
      WRITE (6,32) DA,EVAP,CEVAP,CSEVP,CIRR,RAIN,CDRAIN,CWADD,CPROD,WC(1
     1),WC(2),WC(3),WC(4),WC(5),CPD,C(6),C(7),C(8),C(9),C(10)
      CPROD=CPROD*POTPRO
      CPD=CPD*POTPRO
      WRITE(6,94)CPD,CPROD
      READ (5,1) FIRR
   94 FORMAT('   TOT DRY MATTER=  ',F10.4,' GRAIN =',F10.4)
      WRITE (6,93)FIRR
   93 FORMAT('     FIRR=  ',F4.3)
      DO 92 J=2,IR,2
   92 GIRR(J)=GSAV(J)*FIRR
      DO 53 I=1,5
   53 BGSM(I)=BGSAV(I)
      GO TO 6
    7 STOP
      END
128.        -2.0        0.5        91.        100.       1.0
 28  1 21 54
.4          .4         .0
4.5         5.8        4.2        3.9        2.1
45.         45.        30.        30.        30.
.10         .13        .14        .13        .07
.16         .34        .09        .09        .32
  6 5.80 10 0.20 11 0.79 21 1.63 22 0.23 24 1.04 28 0.33 62 1.27 63 0.13
 64 0.84 79 0.56105 0.30111 0.30129 9.99
    .24.239   9 .62.619 17 .44.430 24 .54.480 31 .69.460 38 .73.380 45 .67.240
 52 .63.150 59 .66.110 66 .72.120 73 .54.090 79 .58.100 86 .60.100093 .58.100
100 .41.080107 .41.080118 .41.090129 .99.990
 35 1.90 42 2.10 49 1.80 55 1.40 62 1.40 70 1.40 76 2.00 82 1.50 89 1.60
 98 1.30129
```

```
      WRITE (6,72)WHC
      READ (5,1)DDST
      WRITE (6,73)DDST
      READ (5,60)(R(J),J=1,N)
      WRITE (6,61)(R(J),J=1,N)
      READ(5,41)(ET(J),J=1,IET)
      DO 97 I=1,IET,3
      ET(I+1)=ET(I+1)*VARY
   97 ET(I+2)=ET(I+2)*VARY
      WRITE (6,42)(ET(J),J=1,IET)
      READ(5,60) (GIRR(J),J=1,IR)
      IF(GIRR(1).GT.499) GO TO 7
      DO 91 J=2,IR,2
   91 GSAV(J)= GIRR(J)
    6 WRITE (6,43) (GIRR(J),J=1,IR)
      IF(NN.EQ.0) WRITE(6,101)
      IF(NN.NE.0) WRITE(6,141)
      KE=1
      CFD=0.0
      KG=1
      SEVN=ET(3)
      EVAP=ET(2)
      I=1
      K=1
      J=1
      DD=DDST(J)
      C1=0.
      CEV=0.
      C2=0.
      C3=0.
      C4=0.
      CSE=0.
      CET=0.0
      SEV=SEVN
      CEVAP=0.0
      CSEVP=0.0
      CWADD=0.0
      RAIN=0.0
      ET(IET+1)=DAYS+2.
      R(N+1)=DAYS+2.
      GIRR(IR+1)=DAYS+2.
      DA=1.
      FRQI=DA
      CDRAIN=0.0
      CPD=0.0
    2 IF(DA.LT.R(K).AND.DA.LT.GIRR(KG)) GO TO 3
      CIRR=0.0
      CPROD=1.0
      WADD=0.
      IF(DA.LT.R(K)) GO TO 50
      K=K+2
      RAIN=RAIN+R(K-1)
      WADD=R(K-1)
      GO TO 51
   50 WADD=GIRR(KG+1)+WADD
      CIRR=CIRR+GIRR(KG+1)
      KG=KG+2
      GO TO 55
   51 IF(DA.EQ.GIRR(KG))GO TO 50
   55 SEV=SEVN
      CWADD=CWADD+WADD
```

```
$ SET NEW SEQ
C      CALCULATION OF SOIL EVAPORATION, TRANSPIRATION AND CROP PRODUCTION
C      DAYS IS NUMBER OF DAYS IN SEASON
C      AIRDRY IS DIFF BETWEEN PERM WILT PT & AIRDRY H2O CONTENT (NEG.)
C      AWFAC IS AVAIL WATER FACTOR--THE FRACTION BELOW WHICH ACTUAL TRANS
C        IS LESS THAN POTENTIAL
C      RTDMAX IS MAXIMUM ROOTING DEPTH
C      N,IR,IET ARE COUNTERS FOR READING IN ARRAYS
C      NN--IF NN=/,DAILY COMPUTATIONS ARE NOT WRITTEN OUT
C      E ARRAY IS GRAIN EXPONENT FOR EACH GROWTH STAGE
C      BGSM ARRAY IS THE BEGINNING SOIL WATER CONTENT IN EACH LAYER
X      THK IS THE THICKNESS OF EACH LAYER
C      WHC IS THE FRACTIONAL AVAIL WATER IN EACH LAYER
C      DDST IS THE FRACTION OF THE SEASON IN EACH GROWTH STAGE
C      R ARRAY CONTAINS THE DAY RAIN OCCURS FOLLOWED BY THE AMOUNT
C      ET IS DAY AT BEGINNING PERIOD FOLLOWED BY ETPOT AND EPOT
C      GIRR IS DAY IRRIGATION OCCURRES FOLLOWED BY AMOUNT
C      SEVN  IS THE POTENTIAL SOIL EVAPORATION RATE
C      EVAP IS EVAPOTRANSPIRATION RATE PER DAY -- POTENTIAL
       DIMENSION E(5),C(10),R(200),GIRR(200),ET(300),GSAV(200)
       DIMENSION BGSM(5),THK(5),WHC(5),WC(5),DDST(5) ,BGSAV(5)
 10    FORMAT (4I6,14F7.4)
   1   FORMAT(7F10.3)
 40    FORMAT (24I3)
 49    FORMAT (1X, 'N,NN,IR,IET: ',24I3)
 30    FORMAT(1H ,'DAYS EVAP TRANS SOLEV IRRIG RAIN DRAIN TOT USE GRAIN
   1      WC1      WC2       WC3      WC4      WC5  DRY MAT   C6        C7        C8
   2C9        C10')
  101 FORMAT(1H ,'DAYS EVAP TRANS SOLEV IRRIG RAIN DRAIN CWADD    WC1
   1     WC2      WC3      WC4      WC5   ETMAX ETDEF  DEPL  ETACT  WATADD')
  141 FORMAT(1H ,'DAYS EVAP TRANS SOLEV IRRIG RAIN DRAIN CWADD   SM1
   1     SM2      SM3      SM4      SM5    KC')
 5     FORMAT(1H ,'E ARRAY',7F10.3)
   32 FORMAT (I4,7F6.2,12F7.3)
 41    FORMAT (7(I3,F4.2,F4.2))
 42    FORMAT (' DAY ETPOT EPOT',10(I3,F4.2,F4.2)/2X11(I3,F4.2,F4.2)/2X11
   1(I3,F4.2,F4.2))
 43    FORMAT(' DAY IRRIGATION', 13(I4,F5.2)/2X 14(I4,F5.2)/2X 14(I4,F5.2
   $)/2X 14(I4,F5.2)/2X 14(I4,F5.2))
   45 FORMAT(  ' DAYS=',I4,' AIRDRY=',F5.2,' AWFAC=',F4.2,'  RTDAMX=',F4
   1.0,' POTPRO=',F6.0,' VARY= ',F5.3)
 60    FORMAT(9(I3,F5.2))
 61    FORMAT (' DAY RAIN', 13(I4,F5.2)/2X 14(I4,F5.2))
   70 FORMAT (1H,'BGSM ARRAY',7F10.3)
   71 FORMAT (1H,'THK  ARRAY',7F10.3)
   72 FORMAT (1H,'WHC  ARRAY',7F10.3)
   73 FORMAT (1H,'DDST ARRAY',7F10.3)
 20    READ (5,1) DAYS,AIRDRY,AWFAC,RTDAMX,POTPRO,VARY
       WRITE (6,45)DAYS,AIRDRY,AWFAC,RTDAMX,POTPRO,VARY
       READ(5,40)N,NN,IR,IET
       WRITE (6,49)N,NN,IR,IET
       READ (5,1) E
       WRITE(6,5) E
       READ(5,1)BGSM
       DO 31 I=1,5
 31    BGSAV(I)=BGSM(I)
       WRITE (6,70)BGSM
       READ (5,1)THK
       WRITE (6,71)THK
       DEPMAX=THK(1)+THK(2)+THK(3)+THK(4)+THK(5)
       READ (5,1)WHC
```

References

BIELORAI, H., ARNON, I., BLUM, A., ELKANA, Y. and REISS, A. (1964). The effects of irrigation and inter-row spacing on grain sorghum production. *Israel J. Agric. Res.* **14**:227–236

BRIGGS, L.J. and SHANTZ, H.L. (1913). The water requirements of plants. II. A review of literature. U.S. Dept. Agric. Bur. Plant Ind. Bull. 285

CHILDS, S.W., GILLEY, J.R. and SPLINTER, W.E. (1977). A simplified model of corn growth under moisture stress. *Trans. ASAE* **20**(5):858–865

CHILDS, S.W. and HANKS, R.J. (1975). Model of soil salinity effects on crop growth. *Soil Sci. Soc. Amer. Proc.* **39**:617–622

DE WIT, C.T. (1958). Transpiration and crop yields. Institute of Biological and Chemical Research on Field Crops and Herbage. Wageningen, The Netherlands, Verse-Landvouwk, onder Z. No. 64. 6–S, Gravenhage

DOORENBOS, J. and PRUITT, W.O. (1977). Guidelines for predicting crop water requirements. FAO Irrigation and Drainage Paper No. 24, Rome, Italy. 144 pp

FISHBACH, P.E. and SOMERHALDER, B.R. (1972). Soil moisture extraction and irrigation design requirements for corn. Paper No. 72-770. Presented at Am. Soc. Agric. Eng. Meet., Chicago, Dec. 11–15

HANKS, R.J. (1974). Model for predicting plant yield as influenced by water use. *Agron. J.* **66**:660–665

HANKS, R.J. and ANDERSEN, J.C. (1978). Physical and economic evaluation of irrigation return flow and salinity on a farm. Chapter 8 in: YARON, D. (ed.), *Salinity in Irrigation and Water Resources*, Marcel Dekker, New York

HANKS, R.J., GARDNER, H.R. and FLORIAN, R.L. (1969). Plant growth-evapotranspiration relations for several crops in the Central Great Plains. *Agron. J.* **60**:30–34

HANWAY, J.J. (1963). Growth stages of corn (*Zea mays*, L.). *Agron. J.* **55**:487–492

HILL, R.W., HANKS, R.J., KELLER, J. and RASMUSSEN, V.P. (1974). Predicting corn growth as affected by water management: An example. CUSUSWASH 211(d)-6. Utah State Univ., Logan, Utah. 18 pp

HILL, R.W., JOHNSON, D.R. and RYAN, K.H. (1979). A model for predicting soybean yields from climatic data. *Agron. J.* **71**:251–256

HILLEL, D. and GURON, Y. (1973). Relation between evapotranspiration rate and maize yield. *Water Resour. Res.* **9**:743–748

JENSEN, M.E. (1968). Water consumption by agricultural plants. In KOZLOWSKI, T.T. (ed.), *Water Deficits and Plant Growth*, Vol. 2, Academic Press, New York.

JENSEN, M.E. (ed.) (1973). Consumptive use of water and irrigation water requirements. Techn. Com. on Irrigation Water Requirements, Irrigation and Drainage Div. ASCE, New York. 215 pp

JOHNSON, D.R., MURPHY, W.J. and LEUDDERS, V.D. (1973a). Soybean varieties and date of planting. 1973 Res. Rep. Southwest Missouri Center, pp. 21–24

JOHNSON, D.R., MURPHY, W.J. and LEUDDERS, V.D. (1973b). Field day reports. North Missouri Center, pp. 13–15

JURY, W.A. and TANNER, C.B. (1975). Advective modifications of the Priestley and Taylor evapotranspiration formula. *Agron. J.* **67**:840–842

KANEMASU, E.T., STONE, L.R. and POWERS, W.L. (1976). Evapotranspiration model tested for soybean and sorghum. *Agron. J.* **68**:569–572

KANEMASU, E.T., RASMUSSEN, V.P. and BAGLEY, J. (1978). Estimating water requirements for corn with a "pocket calculator."

MAJOR, D.J., JOHNSON, D.R., TANNER, J.W. and ANDERSON, I.C. (1975). Effects of daylengths and temperature on soybean development. *Crop Sci.* **15**:174–179

MORGAN, T.H., BIERE, A.W. and KANEMASU, E.T. (1979). A dynamic model of corn yield response to water. (Submitted to *Water Resour. Res.*)

NIMAH, M.N. and HANKS, R.J. (1973). Model for estimating soil water and atmospheric interrelations. *Soil Sci. Soc. Amer. Proc.* **37**:522–532

PRIESTLY, C.H.B. and TAYLOR, R.J. (1972). On the assessment of surface heat flux and evaporation using large scale parameters. *Mon. Weather Rev.* **100**:81–92

RASMUSSEN, V.P. and HANKS, R.J. (1978). Spring wheat model for limited moisture conditions. *Agron. J.* **70**:940–944

SHALHEVET, J., MANTELL, A., BIELORAI, H. and SHIMSHI, D. (1976). *Irrigation of Field and Orchard Crops under Semi-Arid Conditions*. IIIC, Volcani Center, Bet Dagan, Israel. 130 pp

STEWART, J.I. and HAGAN, R.M. (1973). Functions to predict effects of crop water deficits. *J. Irr. Drain. Div., ASCE* **99** (LR4):421–439

STEWART, J.I., HAGAN, R.M. and PRUITT, W.O. (1974). Functions to predict optimal irrigation programs. *J. Irr. Drain. Div., ASCE* **100**(IR2):179–199

STEWART, J.I., MISRA, R.D. and PRUITT, W.O. (1975). Irrigation of corn and grain sorghum with a deficient water supply. *Trans. ASCE* **18**(2):270–280

STEWART, J.I., DANIELSON, R.E., HANKS, R.J., JACKSON, E.B., HAGAN, R.M., PRUITT, W.O., FRANKLIN, W.T. and RILEY, J.P. (1977). *Optimizing Crop Production through Control of Water and Salinity Levels in the Soil*. Utah Water Research Lab. PR 151–1, Logan, Utah. 191 pp

TANNER, C.B. (1967). Measurement of evapotranspiration. *In:* HAGAN, R.M., HAISE, H.R. and EDMINSTER, T.W. (eds.), *Irrigation of Agricultural Lands. Agronomy 11, Amer. Soc. Agron.*, Madison, Wisconsin, pp. 534–574

VAN KEULEN, H. (1975). *Simulation of Water Use and Herbage Growth in Arid Regions*. Center Agr. Pub. Doc., Wageningen, The Netherlands. 192 pp

WOLF, J.K. (1977). The evaluation of a computer model to predict the effects of salinity on crop growth. M.Sc. Thesis, Utah State University, Logan, Utah.